Escape from Pain

The History of the Discovery of Anaesthesia
as Written by Sir James Paget in 1879

Edited by Andrew Sadler

Cover Images:

Left: *Sir James Paget. by Sir Leslie Ward (Spy) 1876. The image appeared on the front of 'Vanity Fair'. On 13th February 1876. Paget wrote to his son Stephen Paget: 'I hope you have seen Vanity Fair, the face seems to be like, the figure is absurdly unlike'.*

Top centre: *Crawford Williamson Long - probably the first person ever to administer an anaesthetic; ether for removal of a neck lump in 1842.*

Top right: *William T G Morton who publically demonstrated an ether anaesthetic at the Massachusetts General Hospital in 1846 and claimed first use. He was unaware of Crawford Long's work.*

Bottom centre: *Gardner Quincy Colton who gave the first ever nitrous oxide anaesthetic to:*

Bottom right: *Horace Wells, the dentist from Hartford Connecticut, whose idea it was. Wells submitted to the extraction of a painful molar carried out by his assistant Dr Riggs while Colton administered the gas.*

Copyright

Escape from Pain - The History of a Discovery and the appendix are public domain.

The remainder is © Andrew Sadler who asserts his moral right to be identified as the author.

For images see the acknowledgements section for sources.

ISBN: 978-1-326-03764-2

SOREJAW

Introduction

The fascinating story of the discovery of anaesthesia, and subsequent practice and controversy, has been the subject of much historical research and writing but this monograph written by Sir James Paget in 1879 is as good an introduction to the tale as any I have come across and bears reproduction to a modern audience in digital and low cost printed medium.

I first heard of Sir James Paget in December 1974 two days before I was to sit the Oral Medicine paper for my Dentistry finals examination. The Dean of the Dental School had written a paper about Paget's disease of bone which Paget had described in 1876 and called 'osteitis deformans'; I thought I had better know about it. A decade later, as a medical student, I became acquainted with Paget's disease of the nipple.

I have taken some liberties in that I have changed the title from the original 'The Escape from Pain - The History of a Discovery' so that it might be more easily discovered on internet searches and I have divided up the text with sub-headings for easy reference. A couple of spelling and punctuation errors in the original have been corrected so that I might not be blamed.

I have added some postscripts of my own: a brief summary of how anaesthesia started in England, the relative merits of chloroform and ether, as well as some reports of a pre-anaesthetic surgery. I have added a short summary of Sir James Paget's life and have suggested three books that I admire for further reading on the subject. A list of many of Sir James' writings is appended.

Andrew Sadler December 2014

Contents

Escape from Pain - The History of a Discovery - by Sir James Paget 1879   7

    Humphry Davy and his discovery of the analgesic properties of nitrous oxide   7

    The dentist Horace Wells and the first use of nitrous oxide as an anaesthetic   8

    Early use of ether as a medication and for amusement   10

    Crawford Long and the first known anaesthetic for surgery   13

    Charles Jackson, William Morton and the second discovery of ether anaesthesia   14

    James Simpson and the discovery of chloroform anaesthesia   17

    Colton and the popularisation of nitrous oxide for dental anaesthesia   18

    What became of the great innovators of anaesthesia?   20

    The arguments over approbation   23

    The practical difficulties of the discovery   27

Postscript 1   The Start of Anaesthesia in Britain   31

Postscript 2.   Chloroform and Ether Anaesthesia   35

Postscript 3.   Surgery before Anaesthesia   37

    'Cries from the Operating Theatre'   37

    'Such acute pain that he fainted'   37

    'The patient is then tied up.'   39

    Too many spectators - operation postponed until Saturday   40

    A death on the operating table - nurses and patients on the ward shed tears   41

    Desperate and ineffective methods to reduce pain   43

Postscript 4.   A Short Life of James Paget   45
   Birth and Education   45
   Apprentice to a surgeon in Yarmouth 1830-1834   46
   Pupillage at St. Bartholomew's Hospital 1834-1835   47
   Waiting for suitable employment 1836-43   50
   Teaching appointments at Bart's Hospital 1843-1851   50
   Private Practice 1851 onwards   52
   Medical Politics   53
   Controversy   54
   Old Age   54

Appendix: List of chief writings of James Paget   55

Further Reading   60

Acknowledgements   61

Escape from Pain - The History of a Discovery - by Sir James Paget 1879

The history of the discovery of methods for the prevention of pain in surgical operations deserved to be considered by all who study either the means by which knowledge is advanced or the lives of those by whom beneficial discoveries are made. And this history may best be traced in the events which led to and followed the use of nitrous oxide gas, of sulphuric ether, and of chloroform as anaesthetics – that is, as means by which complete insensibility may be safely produced and so long maintained that a surgical operation, of whatever severity and however prolonged, may be absolutely painless.

Humphry Davy and his discovery of the analgesic properties of nitrous oxide

In 1798, Mr. Humphry Davy, an apprentice to Mr. Borlase, a surgeon at Bodmin, had so distinguished himself by zeal and power in the study of chemistry and natural philosophy, that he was invited by Dr. Beddoes, of Bristol, to become the 'superintendent of the Pneumatic Institution which had been established at Clifton for the purpose of trying the medicinal effects of different gases.' He obtained release from his apprenticeship, accepted the appointment, and devoted himself to the study of gases, not only in their medicinal effects, but much more in all their chemical and physical relations. After two years' work he published his *Researches, Chemical and Philosophical, chiefly concerning Nitrous Oxide*, an essay proving a truly marvellous ingenuity, patience, and courage in experiments, and such a power of observing and of thinking as has rarely if ever been surpassed by any scientific man of Davy's age; for he was then only twenty-two.

In his inhalations of the nitrous oxide gas he observed all the phenomena of mental excitement, of exalted imagination, enthusiasm, merriment, restlessness, from which it gained its popular name of 'laughing gas', and he saw people made, at least for some short time and in some measure, insensible by it. So, among other suggestions or guesses about probable medicinal uses of inhalation of gases, he wrote, near the end of his essay: 'As nitrous oxide in its extensive operation appears capable of destroying physical pain, it may probably be used with advantage during surgical operations in which no great effusion of blood takes place.'

The dentist Horace Wells and the first use of nitrous oxide as an anaesthetic

It seems strange that no one caught at a suggestion such as this. True, the evidence on which it was founded was very slight; it was with a rare scientific power that Davy had thought out so far beyond his facts; but he had thought clearly, and as clearly told his belief. Yet no one earnestly regarded it. The nitrous oxide might have been of as little general interest as the carbonic or any other, had it not been for the strange and various excitements produced by its inhalation. These made it a favourite subject with chemical lecturers, and year after year, in nearly every chemical theatre, it was fun to inhale it after the lecture on the gaseous compounds of nitrogen, and among those who inhaled it there must have been many who, in their intoxication, received sharp and heavy blows, but, at the time, felt no pain. And this went on for more than forty years, exciting nothing worthy to be called thought or observation, till, in December 1844,

Horace Wells

Mr. Colton, a popular itinerant lecturer on chemistry, delivered a lecture on 'laughing gas' in Hartford, Connecticut. Among his auditors was Mr. Horace Wells, an enterprising dentist in that town, a man of some power in mechanical invention. After the lecture came the usual amusement of inhaling the gas, and Wells, in whom long wishing had bred a kind of belief that something might be found to make tooth-drawing painless, observed that one of the men excited by the gas was not conscious of hurting himself when he fell on the benches and bruised and cut his knees. Even when he became calm and clear-headed the man was sure that he did not feel pain at the time of his fall. Wells was at once convinced – more easily convinced than a man of more scientific mind would have been - that, during similar insensibility, in a state of intense nervous excitement, teeth might be drawn without pain, and he determined that himself and one of his own largest teeth should be the first for trial. Next morning Colton gave him the gas, and his friend Dr. Riggs extracted his tooth. He remained unconscious for a few moments, and then exclaimed, 'A new era in tooth-pulling! It did not hurt me more than the prick of a pin. It is the greatest discovery every made.'

In the next three weeks Wells extracted teeth from some twelve or fifteen persons under the influence of the nitrous oxide, and gave pain to only two or three. Dr. Riggs, also, used it with the same success, and the practice was well known and talked of in Hartford.

Encouraged by his success Wells went to Boston, wishing to enlarge the reputation of his discovery and to have an opportunity of giving the gas to some one undergoing a surgical operation. Dr. J. C. Warren, the senior Surgeon of the Massachusetts General Hospital, to whom he applied for this purpose, asked him to show first its effects on some one from whom he would draw a tooth. He undertook to do this in the theatre of the medical college before a large class of students, to whom he had, on a previous day, explained his plan. Unluckily, the bag of gas from which the patient was inhaling was taken away too soon; he cried out when his tooth

Nitrous oxide party 1840s

was drawn; the students hissed and hooted; and the discovery was denounced as an imposture.

Wells left Boston disappointed and disheartened; he fell ill, and was for many months unable to practise his profession. Soon afterwards he gave up dentistry, and neglected the use and study of the nitrous oxide, till he was recalled to it by a discovery even more important than his own.

The thread of the history of nitrous oxide may be broken here.

Early use of ether as a medication and for amusement

The inhalation of sulphuric ether was often, even in the last century, used for the relief of spasmodic asthma, phthisis, and some other diseases of the chest. Dr. Beddoes and others thus wrote of it; but its utility was not great, and there is no evidence that this use of it had any influence on the discovery of its higher value, unless it were, very indirectly, in its having led to its being found useful for soothing the irritation produced by inhaling chlorine. Much more was due to its being used, like nitrous oxide, for the

Patient.—"This is really quite delightful—a most beautiful dream."

From Punch 1847

fun of the excitement which its diluted vapour would produce in those who freely inhaled it.

The beginning of its use for this purpose is not clear. In the *Journal of Science and the Arts*, published in 1818 at the Royal Institution, there is a short anonymous statement among the 'Miscellanea', in which it is said, 'When the vapour of ether mixed with common air is inhaled, it produces effects very similar to those occasioned by nitrous oxide.' The method of inhaling and its effects are described, and then 'it is necessary to use caution in making experiments of this kind. By the imprudent inspiration of ether a gentleman was thrown into a very lethargic state, which continued with occasional periods of intermission for more than thirty hours, and a great depression of spirits; for many days the

Crawford Long

pulse was so much lowered that considerable fears were entertained for his life.'

The statement of these facts has been ascribed to Faraday, under whose management the journal was at that time published. But, whoever wrote or whoever may have read the statement, it was, for all useful purposes, as much neglected as was Davy's suggestion of the utility of the nitrous oxide. The last sentence, quoted as it was by Pereira and others writing on the uses of ether, excited much more fear of death than hope of ease from ether-inhalation. Such effects as are described in it are of exceeding rarity; their danger was greatly over-estimated; but the account of them was enough to discourage all useful research.

But, as the sulphuric ether would 'produce effects very similar to those occasioned by nitrous oxide', and was much the more easy to procure, it came to be often inhaled, for amusement, by chemists' lads and by pupils in the dispensaries of surgeons. It was often thus used by young people in many places in the United States. They had what they called 'ether-frolics', in which they inhaled ether till they became merry, or in some other way absurdly excited or, sometimes, completely insensible.

Crawford Long and the first known anaesthetic for surgery

Among those who had joined in these ether-frolics was Dr. Wilhite of Anderson, South Carolina. In one of them, in 1839, when nearly all of the party had been inhaling and some had been laughing, some crying, some fighting – just as they might have done if they had had the nitrous oxide gas – Wilhite, then a lad of seventeen, saw a negro-boy at the door and tried to persuade him to inhale. He refused and resisted all attempts to make him do it, till they seized him, held him down, and kept a handkerchief wet with ether close over his mouth. Presently his struggles ceased; he lay insensible, snoring, past all arousing; he seemed to be dying. And thus he lay for an hour, till medical help came and, with shaking, slapping, and cold splashing, he was awakened and suffered no harm.

The fright at having, it was supposed, so nearly killed the boy, put an end to ether-frolics in that neighbourhood; but in 1842, Wilhite had become a pupil of Dr. Crawford Long, practising at that time at Jefferson (Jackson County, Georgia). Here he and Dr. Long and three fellow-pupils often amused themselves with the ether inhalation, and Dr. Long observed that when he became furiously excited, as he often did, he was unconscious of the blows which he, by chance, received as he rushed or tumbled about. He observed the same in his pupils; and thinking over this, and emboldened by what Mr. Wilhite told him of the negro-boy recovering after an hour's insensibility, he determined to try whether the ether-inhalation would make anyone insensible of the pain of an operation. So, in March 1842, nearly three years before Wells's observations with the nitrous oxide, he induced a Mr. Venable, who had been very fond of inhaling ether, to inhale it till he was quite insensible. Then he dissected a tumour from his neck; no pain was felt, and no harm followed. Three months later, he similarly removed another tumour from him; and again, in 1842 and in 1845, he operated on another three patients, and none felt pain. His operations were known and talked of in his

William Morton

neighbourhood; but the neighbourhood was only that of an obscure little town, and he did not publish any of his observations. The record of his first operation was only entered in his ledger:

'James Venable, 1842. Ether and excising tumour, $2.00'.

He waited to test the ether more thoroughly in some greater operation than those in which he had yet tried it, and then he would have published his account of it. While he was waiting, others began to stir more actively in busier places, where his work was quite unknown, not even heard of.

Charles Jackson, William Morton and the second discovery of ether anaesthesia

Among those with whom, in his unlucky visit to Boston, Wells talked of his use of the nitrous oxide, and of the great discovery which he believed that he had made, were Dr. Morton and Dr. Charles Jackson, men widely different in character and pursuit, but inseparable in the next chapter of the history of anaesthetics.

Morton was a restless energetic dentist, a rough man, resolute to get practice and make his fortune. Jackson was a quiet scientific gentleman, unpractical and unselfish, in good repute as a chemist, geologist and mineralogist. At the time of Wells's visit, Morton, who had been his pupil in 1842, and for a short time, in 1843, his partner, was studying medicine and anatomy at the Massachusetts

Medical College, and was living in Jackson's house. Neither Morton nor Jackson put much if any faith in Wells's story, and Morton witnessed his failure in the medical theatre. Still, Morton had it in his head that tooth-drawing might somehow be made painless, and even after Wells had retired from practice, he talked with him about it, and made some experiments, but, having no scientific skill or knowledge, they led to nothing. Still, he would not rest, and he was guided to success by Jackson, whom Wells advised him to ask to make some nitrous oxide gas for him.

Jackson had long known, as many others did, of sulphuric ether being inhaled for amusement and of its producing effects like those of nitrous oxide: he knew also of its employment as a remedy for the irritation caused by inhaling chlorine. He had himself used it for this purpose, and once, in 1842, while using it, he became completely insensible. He had thus been led to think that the pure ether might be used for the prevention of pain in surgical operations; he spoke of it with some scientific friends, and sometimes advised a trial of it; but he did not urge it or take any active steps to promote even the trial. One evening, Morton, who was now in practice as a dentist, called on him, full of some scheme which he did not divulge, and urgent for success in painless tooth-drawing. Jackson advised him to use the ether, and taught him how to use it.

On that same evening, the 30th of September, 1846, Morton inhaled the ether, put himself to sleep, and, when he awoke, found that he had been asleep for eight minutes. Instantly, as he tells, he looked for an opportunity of giving it to a patient; and one just then coming in, a stout healthy man, he induced him to inhale, made him quite insensible, and drew his tooth without his having the least consciousness of what was done.

But the great step had yet to be made – the step which Wells would have tried to make if his test-experiment had not failed. Clearly, operations as swift as that of tooth-drawing might be rendered painless, but could it be right to incur the risk of

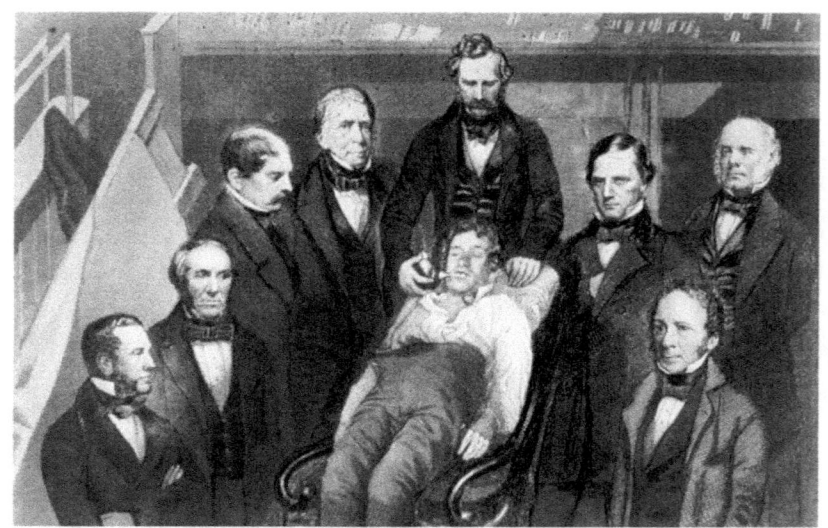

William Morton administers ether at Massachusetts General Hospital October 16th 1846

insensibility long enough and deep enough for a large surgical operation? It was generally believed that in such insensibility there was serious danger to life. Was it really so? Jackson advised Morton to ask Dr. J. C. Warren to let him try, and Warren dared to let him. It is hard, now, to think how bold the enterprise must have seemed to those who were capable of thinking accurately on the facts then known.

The first trial was made on the 16th of October, 1846. Morton gave the ether to a patient in the Massachusetts General Hospital and Dr. Warren removed a tumour from his neck. The result was not complete success; the patient hardly felt the pain of the cutting, but he was aware that the operation was being performed. On the next day, in a severer operation by Dr. Hayward, the success was perfect; the patient felt nothing, and in long insensibility there was no appearance of danger to life.

James Simpson

The discovery might already be deemed complete; for the trials of the next following days had the same success, and thence onwards the use of the ether extended over constantly widening fields. A coarse but feeble opposition was raised by some American dentists; a few surgeons were over-cautious in their warnings against suspected dangers; a few maintained that pain was very useful, necessary perhaps to sound healing; some were hindered by their dislike of the patent which Morton and Jackson took out; but as fast as the news could be carried from one continent to another, and from town to town, so fast did the use of ether spread. It might almost be said that in every place, at least in Europe, where the discovery was promoted more quickly than in America, the month might be named before which all operative surgery was agonising, and after which it was painless.

James Simpson and the discovery of chloroform anaesthesia

But there were other great pains yet to be prevented, the pains of childbirth. For escape from these the honour and deep gratitude are due to Sir James Simpson. No energy, or knowledge, or power of language less than his could have overcome the fears that the insensibility, which was proved to be harmless in surgical operations and their consequences, should be often fatal or very mischievous in parturition. And to these fears were added a crowd of pious protests (raised, for the most part, by men) against so gross an interference as this seemed with the ordained course of human nature. Simpson, with equal force of words and work, beat all down; and by his adoption of chloroform as a substitute for ether promoted the whole use of anaesthetics.

Colton and the popularisation of nitrous oxide for dental anaesthesia

Ether and chloroform seemed to supply all that could be wished from anaesthetics. The range of their utility extended; the only question was as to their respective advantages, a question still unsettled. Their potency was found absolute, their safety very nearly complete, and, after the death of Wells in 1848, nitrous oxide was soon neglected and almost forgotten. Thus it remained till 1862, nearly seventeen years, when Mr. Colton, who still continued lecturing and giving the gas 'for fun', was at New Haven, Connecticut. He had often told what Wells had done with nitrous oxide at Hartford, and he wanted other dentists to use it, but none seemed to care for it till, at New Britain, Dr. Dunham asked him to give it to a patient to whom it was thought the ether might be dangerous. The result was excellent, and in 1863 Dr. Smith of New Haven substituted the nitrous oxide for ether in his practice and used it very frequently. In the nine months following his first use of it, he extracted without pain nearly 4,000 teeth. Colton, in the following year, associated himself with a dentist in New York and established the Colton Dental Association, where the gas was given to many thousands more. Still, its use was very slowly admitted. Some called it dangerous, others were content with chloroform and ether, others said that the short pangs of tooth-drawing had better be endured. But in 1867 Mr. Colton came to Paris and Dr. Evans at once promoted his plan. In 1868 he came to London and, after careful study of it at the Dental Hospital, the nitrous oxide was speedily adopted, both by dentists and by the administrators of anaesthetics. By this time it had saved hundreds of thousands of people from the sharp pains of all kinds of

Gardner Quincy Colton

operations on the teeth and of a great number of the surgical operations that can be quickly done.

Such is the history of the discovery of the use of anaesthetics. Probably, none has ever added so largely to that part of happiness which consists in the escape from pain. Past all counting is the sum of happiness enjoyed by the millions who, in the last three-and-thirty years have escaped the pains that were inevitable in surgical operations; pains made more terrible by apprehension, more keen by close attention; sometimes awful in a swift agony, sometimes prolonged beyond even the most patient endurance, and then renewed in memory and terrible in dreams. These will never be felt again. But the value of the discovery is not limited by the abolition of these pains or the pains of childbirth. It would need a long essay to tell how it has enlarged the field of useful surgery, making many things easy that were difficult, many safe that were too perilous, many practicable that were nearly impossible. And, yet more variously, the discovery has brought happiness in the relief of some of the intensest pains of sickness, in quieting convulsion, in helping to the discrimination of obscure diseases. The tale of its utility would not end here; another essay might tell its multiform uses in the study of psychology, reaching even to that of the elemental processes in plants, for these, as Claude Bernard has shown, may be completely for a time suspended in the sleep produced by chloroform or ether.

What became of the great innovators of anaesthesia?

And now, what of the discoverers?[1] What did time bring to those who brought so great happiness to mankind?

Long, out-stripped in the race of discovery, continued in his quiet useful life in practice at Jefferson, and after 1850 at Athens (Georgia). The fact of his discovery was not known beyond his own narrow circle of friends till the use of both ether and chloroform had become general. Then his claim to honour was as little heard as one gentle voice might be in the uproar of a confused and noisy crowd. In 1853, when Morton made one of his claims on Congress for a large reward in money, it was shown that Long had used the ether more than four years before him. The claim to honour could not be denied. It was admitted by Jackson, who wrote that if Long 'had written to him in season' he 'would have presented his claims to the Academy of Sciences of France.' But nothing followed; and Long's name and worth were known to very few till, in 1877, Dr. Marion Sims published a full account of his observations. Long was then an old man, still at work in his profession and obliged to work very hard, for he had suffered heavy losses in the Civil War. He had the esteem of all who knew him; he was, as Jackson described him, 'a very modest, retiring man, an

---

[1] *Those only are here reckoned as discoverers from whose work may be traced not merely what might have been the beginning of the discovery, but the continuous history of events consequent on the evidence of its truth. Long, it is true, might under this rule be excluded; yet his work cannot fairly be separated from the history. Of course in this, as in every similar case, there were some who maintained that there was nothing new in it. Before 1842 there were many instances in which persons underwent operations during insensibility. There may be very reasonable doubts about what is told of the ancient uses of Indian hemp and mandragora; but most of those who saw much surgery before 1846 must have seen operations done on patients during insensibility produced by narcotics, dead-drunkenness, mesmerism, large losses of blood or other uncertain and often impracticable methods. Besides, there were many guesses and suggestions for making operations painless. But they were all fruitless; and they fail at that which may be a fair test for most of the claims of discoverers – the test of consequent and continuous history. When honour is claimed for the authors of such fruitless works as these, it may fairly be said that blame rather than praise is due to them. Having seen so far as they profess, they should not have rested till they could see much further.*

honourable man in all respects.' Last year he died, and then he was honoured by his portrait being presented by Mr. Stuart of New York to the Alumni of the University of Georgia and placed in the capital of that State.

Of Wells, it has already been told how, after his failure in the attempt to show the value of nitrous oxide at Boston, he went home disheartened, and was long ill and unable to practise his profession. He gave up his dentistry and occupied himself in many things, the last of them picture-dealing, and he left the study of the nitrous oxide till after the full discovery of the use of ether. Then he tried in vain to prove that his method of anaesthesia was the best and safest, not in dentistry alone, but in severer surgery; he and Dr. Marcy, he said, had used ether before Morton did, and found it in no way better than the nitrous oxide. But it was in vain that he tried to gain honour or reward for priority of discovery, or to recover the position and the practice he had given up. Constant disappointment weighed heavily on him, for he was a gentle, sensitive, enthusiastic man. At last, weary and wretched, he became insane, and in 1848 committed suicide. Some twenty years afterwards, the nitrous oxide was in full use; his statue was set up in Hartford; and five years later, his widow, still in poverty, was helped by a subscription.

Morton, as vainly, but with longer contest, strove to 'make his fortune'. As soon as the value of the ether-inhalation was proved, he took out a patent for its use, and Jackson was induced to join him in this wrong. But at once there arose a fierce, coarse controversy as to which of the two should have the honour of priority of discovery, and what should be their several profits from the patent. The principals did not write so much as their friends and their attorneys; these were vehement and profuse, and the dispute was made yet more coarse and bitter by those who acquired money-interest in the patent, and by the editors of newspapers in which Morton advertised for practice and for the sale of licences to use his patent. One cannot read the controversy without utter shame at the degradation of truths which should have been told only in the pure language of science and humanity. Some of it is so written; but more is foul with conflicting affidavits, special pleadings of lawyers, perversions of low journalists and mutual charges of falsehood, fraud, and imposition. Of course, no good came of it: Jackson retired from it as soon as he could, and Morton found his patent useless. He and his friends often spoke of the ether as a new compound, and some called it Letheon; and he set up a Letheon-establishment, but every one knew that the 'compound' was sulphuric ether. Jackson never attempted to conceal it, and there was no need of apparatus in its use. The patent cost so much more than it yielded that in a year or two Morton was a very poor man, needing money-help from his friends. A thousand dollars were subscribed for him, and then there was a bitter quarrel as to whether they were given for honour or for charity.

Many times he petitioned Congress for some large reward; he spent time and money in 'lobbying' and the worse means of gaining votes, but all was in vain. His petitions were met with protests from Jackson, with claims for Wells's family, or for Long, and every discussion raised the old controversy, and made it hotter with the heat of new personal and political animosities; for the South was then dominant and Morton was a Northerner. At the end of every attempt to get money he found himself rather poorer, in less practice, and in worse repute; people became utterly tired of the quarrel and, for the most part, indifferent to the claims of all who were engaged in it. Even the discovery itself was less esteemed in America than in Europe, so that that could be often repeated which Dr. Hayward had said of it: 'The only spot in Christendom in which the discovery was received with coldness... was in our own country.'

Thus Morton worried his way through life for two-and-twenty years. His friends helped him from time to time with money and encouragement, but at last even his rough hard nature yielded. After one of his many failures to obtain money from Congress, driven half wild by poverty and disappointment, he returned from

Washington to New York in a state of intense mental excitement. He was kindly tended by Dr. Lewis Sayre and carefully watched. One day, in July 1868, he drove out with his attendant, wanting, as he said, fresh air 'to cool his brain.' Returning home he suddenly jumped from the carriage, ran through a plantation to the border of an adjacent lake, put his head into the water 'to cool his brain', was found insensible, and in a few hours died.

Jackson found himself in trouble as soon as his belief that ether might be used to produce insensibility in surgical operations was proved true by Morton. He seems to have had no selfish view beyond that of maintaining his claim to the honour of the discovery; but to maintain this, he was involved in the discredit of the patent, and in all the controversies had to endure coarse abuse and the imputation of trickery and low motives. He separated himself from Morton as soon as he could, took as little part as possible in the controversy, and all that he wrote was gentle and courteous in comparison with the language of his opponents. He held on his course in the study of mineralogy and geology, wrote many good papers on them, was well esteemed in all the scientific societies of his country, and received scientific honours from abroad. But some years ago his mental power failed, and now he is in an asylum, without either wealth or honours, but happy in the peaceful enjoyment of genial and benevolent delusions.

Simpson had a more prosperous career than these. His introduction of the use of anaesthetics into obstetric practice, and the generally wider range for their employment which was due to his adoption of chloroform in place of ether, were part of the merits which gained for him, among many other honours, a baronetcy and a place among the most successful physicians of his time, a statute in Edinburgh and a memorial bust in Westminster Abbey.

The arguments over approbation

It is hard to repress a first feeling of shame and anger that the American discoverers, great benefactors as they were, should have

Wells in Bushnell Park Hartford, an inscription reads:

HORACE WELLS
THE DISCOVERER OF
ANAESTHESIA

DECEMBER 1844

been left by their fellow-countrymen to die poor, without honour, scarcely thanked for their work. The whole world owes to them immeasurable happiness. America owes them, besides, the honour of a great national renown. Where, then, was the bounty which, in that country, in so many instances, has been splendid in its grandeur? Where the keen jealousy for national honour? It looks as if all had fallen into some sullen ingratitude and indifference.

But, if we look more deeply, we may find no reason for blaming the American people; rather, we may find that they did only what, in the like conditions, would have been done by ourselves or any others. The case was one in which it was made necessary to satisfy, if possible, both gratitude and justice. This might have been easy if there had been only one claimant; gratitude might have been profuse and bountiful, and justice might have approved or, at least, been silent. But there were never less than two, and generally four, claimants; and where gratitude might with a free hand have been ready to give honours and rewards to them all, justice was invoked that the gifts might be in measure proportioned to their several merits. Then gratitude, waiting on justice, became irresolute and cold, or was distracted by new objects.

It may seem very hard; but let any one or, much better, let any four or five, as if sitting in council, think what they would have done; how they would have satisfied at once their gratitude and their sense of justice; how much, suppose, of any great vote by Congress they would have given to each claimant?

How much should Long have had? He first used a true, safe, and sufficient anaesthetic in surgery, and used it with such success that, if he had quickly published his facts, he could not but have been regarded as the great discoverer. It was the fault of his position more than of himself that his facts were not sooner known; and for his delay he might, in so grave a case, plead prudence. But was he then to have no reward?

And what should have been Wells's share? He certainly discovered the use of nitrous oxide, and from his success with it may be traced, not only the knowledge of its whole present utility, but the continuous history of the complete discovery of anaesthetics. True, he soon left the field, disheartened and as if in distrust of his own work; but before he left it he had set Morton on the track, and had thus contributed to the discovery of the uses of ether and chloroform. These, surely, were great merits; what should have been their reward?

Jackson's claims were of a different kind. He had what may be called a scientific idea of the anaesthetic use of ether, but he gave it no active life, no clear persuasive expression. His mind was chiefly occupied in fields of science far apart from active surgery; the great idea needed transplantation. But, when we see to what it grew, we must admit that he who bred and nurtured it, and then gave it to be planted, had great claims to honour.

Morton answered well to the definition given, it is said, by Sydney Smith: 'He is not the inventor who first says the thing, but he who says it so long, loudly, and clearly, that he compels mankind to hear him.' Without either skill, or knowledge, or ingenuity, he supplied the qualities without which the complete discovery of anaesthetics might have been, at least, long delayed – boldness, perseverance, self-confidence. While Long waited, and Wells turned back, and Jackson was thinking, and those to whom they had talked were neither acting nor thinking, Morton, the 'practical man', went to work and worked resolutely. He gave ether successfully in severe surgical operations, he loudly proclaimed his deeds, and 'compelled mankind to hear him.' His claim was very clear.

Probably, most people would agree that all four deserved reward; but that which the controversy and the patent and the employment of legal advisers made it necessary to determine was, whether more than one deserved reward and, if more than one, the proportion to be assigned to each. Here was the difficulty. The French Academy of Sciences in 1850 granted equal shares in the Monthyon Prize to Jackson and to Morton; but Long was unknown to them, and, at the time of the award, the value of nitrous oxide was so hidden by the greater value of ether that Wells's claim was set aside. A memorial column was erected at Boston, soon after Morton's death in 1868, and here the difficulty was shirked by dedicating the column to the discovery of ether, and not naming the discoverers. The difficulty could not be thus settled; and, in all probability, our supposed council of four or five would not solve it. One would prefer the claims of absolute priority; another those of suggestive science; another the courage of bold adventure; sentiment and sympathy would variously affect their judgements. And if we suppose that they, like the American Congress, had to discuss their differences within sound of such controversies as followed Morton's first use of ether, or during a war of pamphlets, or under burdens of parliamentary papers, we should expect that their clearest decision would be that a just decision could not be given, and that gratitude must die if it had to wait till distributive justice could be satisfied. The gloomy fate of the American discoverers makes one wish that gratitude could have been let flow of its own impulse; it would have done less wrong than the desire for justice did. A lesson of the whole story is that gratitude and justice are often incompatible; and that when they conflict, then, usually, *summum jus summa injuria*.

The practical difficulties of the discovery

Another lesson, which has been taught in the history of many other discoveries, is clear in this – the lesson that great truths may be very near us and yet be not discerned. Of course, the way to the discovery of anaesthetics was much more difficult than it now seems. It was very difficult to produce complete insensibility with nitrous oxide till it could be given undiluted and unmixed; this required much better apparatus than Davy or Wells had; and it was hardly possible to make such apparatus till india-rubber manufacturers were improved. It was very difficult to believe that profound and long insensibility could be safe, or that the appearances of impending death were altogether fallacious. Bold as Davy was, bold even to recklessness in his experiments on himself, he would not have ventured to produce deliberately in any one a state so like a final suffocation as we now look at unmoved. It was a boldness not of knowledge that first made light of such signs of dying, and found that what looked like a sleep of death was as safe as the beginning of a night's rest. Still, with all fair allowance for these and other difficulties, we cannot but see and wonder that for more than forty years of this century a great truth lay unobserved, though it was covered with only so thin a veil that a careful physiological research must have discovered it. The discovery ought to have been made by following the suggestion of Davy. The book in which he wrote that 'nitrous oxide – capable of destroying physical pain – may probably be used with advantage during surgical operations', was widely read, and it would be hard to name a man of science more widely known and talked of than he was. Within two years of the publication of his *Researches*, he was appointed to a professorship in the Royal Institution; and in the next year he was a favourite in the fashionable as well as in the scientific world; and all his life through he was intimately associated with those among whom all the various motives for desiring to find some means 'capable of destroying physical pain' would be most strongly felt. Curiosity, the love of truth, the love of marvels, the desire of ease, self-interest, benevolence – all were alert in the minds of men and women who knew and trusted whatever Davy said or wrote, but not one mind was earnestly directed to the rare promise which his words contained. His own mind was turned with its full force to other studies; the interest in surgery which he may have felt during his apprenticeship at Bodmin was lost in his devotion to poetry, philosophy, and natural science, and there is no evidence that he urged others to undertake the study which he left. Even his biographers, his brother Dr. John Davy, and his intimate friend Dr. Paris, both of whom were very capable physicians and men of active intellect, say nothing of his suggestion of the use of nitrous oxide. It was overlooked and utterly forgotten till the prophecy was fulfilled by those who had never heard of it. The same may be said of what Faraday, if it were he, wrote of the influence of sulphuric ether. All was soon forgotten, and the clue to the discovery, which would have been far easier with ether than with nitrous oxide, for it needed no apparatus, and even required mixture with air, was again lost. One could have wished that the honour of bringing so great a boon to men, and so great a help in the pursuit of knowledge, had been won by some of those who were giving themselves with careful cultivation to the search for truth as for its own sake. But it was not so: science was utterly at fault; and it was shown that in the search for truth there are contingencies in which men of ready belief and rough enterprise, seeking for mere utility even with selfish purposes, can achieve more than those who restrain themselves within the range of what seems reasonable.

Such instances of delay in the discovery of truth are always wondered at, but they are not uncommon. Long before Jenner demonstrated the utility of vaccination it was known in Gloucestershire that they who had had cow-pox could not catch the small-pox. For some years before the invention of electric telegraphy, Professor Cumming of Cambridge, when describing to his class the then recent discovery by Oersted of the power of an electric current to deflect a magnet, used to say, 'Here, then, are the elements which would excellently serve for a system of telegraphy.' Yet none of his hearers, active and cultivated as they were, were moved from the routine of study. Laennec quotes a sentence from Hippocrates which, if it had been worthily studied, might have lead to the full discovery of auscultation. Thus it often has been; and few prophecies can be safer than that our successors will wonder at us as we do at those before us; will wonder that we did not discern the great truths which they will say were all around us, within reach of any clear, earnest mind.

They will wonder, too, as we may, when we study the history of the discovery of anaesthetics, at the quietude with which habitual miseries are borne; at the very faint impulse to action which is given by even great necessities when they are habitual. Thinking of the pain of surgical operations, one would think that men would have rushed after the barest change of putting an end to it as they would have rushed to escape from starving. But it was not so; the misery was so frequent, so nearly customary, deemed so inevitable, that, though it excited horror when it was talked of, it did not excite to strenuous action. Remedies were wished for and sometimes tried, but all was done vaguely and faintly; there was neither hope enough to excite intense desire, nor desire enough to encourage hope; the misery was 'put up with' just as we now put up with typhoid fever and sea-sickness, with local floods and droughts, with the waste of health and wealth in the pollutions of rivers, with hideous noises and foul smells, and many other miseries. Our successors, when they have remedied or prevented them, will look back on them with horror, and on us with wonder and contempt for what they will call our idleness or blindness or indifference to suffering.

Postscript 1 The Start of Anaesthesia in Britain

The news of Morton's anaesthetic in Boston soon reached London. A Dr Bigelow of Boston wrote to Dr Francis Boott, a Dental Surgeon of Gower Street, London, about its use by Morton for dental extraction.

Boott administered ether to a girl for the extraction of a tooth in December 1846. He reported that the extraction was performed 'without the least sense of pain or movement of a muscle'. It was later claimed that this event had been pre-dated by the use of ether in Scotland at Dumfries.

Boott informed Robert Liston, a well-known surgeon and Professor of Surgery at nearby University College Hospital. There was a patient in the wards under the care of Dr William Squire, a recently appointed dresser, who was 'worn with exhausting discharge and night suffering occasioned by a disorganised knee joint'. The patient did not have the resolution to submit to the necessary operation of amputation and proclaimed: 'I will die with it (the leg) on.'

Liston doubted that ether would be sufficient to make a patient insensible to the pain of a capital operation such as an amputation even if it was sufficient for a tooth extraction. However on Saturday December 19th he told Squire to obtain some of the vapour and try it upon himself, which he did. On the following Monday Squire administered ether through an apparatus to the patient while Liston amputated the thigh in 24 seconds (they did not tend to count the time taken to secure the vessels, suture or dress the wound). Nevertheless Liston was known as a rapid surgeon, as was necessary in the pre-anaesthetic era.

Liston wrote to Boott:

My Dear Sir,- I tried the ether inhalation today in a case of amputation of the thigh, and in another requiring evulsion of both sides of a great toe-nail, one of the most painful operations in surgery, and with the most perfect and satisfactory results.

It is a very great matter to be able thus to destroy sensibility to such an extent, and without, apparently, any bad result. It is a fine thing for any operating surgeons, and I thank you most sincerely for the information you were so kind as to give me of it.

Yours faithfully, Robert Liston.

Liston was enthused by this 'Yankee dodge', as he called it. Ether was soon in more general, if not yet widespread, use for facilitating painless surgery. There were many enthusiastic reports of it in the Lancet and Provincial Medical and Surgical Journal.

Meanwhile Morton, who had attempted to disguise the ether as 'Letheon' and attempted to patent it, had appointed a Mr James Dorr as his agent in London to enforce and collect revenues for his patent. The prevailing attitude to this was expressed by the editor of the Provincial Medical and Surgical Journal who wrote:

'We question ... whether such a patent can be sustained, and certainly it ought not to be so, any particular form of apparatus may without doubt, be made the property of the inventor; but the attempt to place restrictions on the mode of using a known agent by qualified medical practitioners, is as absurd, as its success would be mischievous.'

Although the news of ether's property of producing insensibility to pain (the term anaesthesia was not yet used) its use was by no means routine. For a start surgery was only carried out as a last resort in attempts to save life so that even Robert Liston only carried out a few operations each month. There was suspicion that the ether might depress the system and make the patient weaker and more likely to succumb and ether was variable and unpredictable in its effects. Some patients were made insensible for surgery; others benefited little. Some patients cried out apparently in pain but had no memory of the procedure afterwards.

The vapour was unpleasant to inhale and many were sick. Many in the church had religious objections.

Probably the first person in Britain to be considered an 'anaesthetist' was John Snow. Snow was a medical practitioner in Soho and was later to receive belated recognition for identifying the source of the cholera epidemic of 1854 as a contaminated water pump in Broad Street rather than 'miasma' or contaminated air as hitherto considered. Snow approached the inhalation of ether in a scientific manner; he started experimenting in 1847 and worked out the dose of ether that should be given. He understood that the temperature of the liquid was important so that the vapour was properly inhaled rather than condensing and he designed various inhalers so the patients could be given a measured dose of the agent.

At St Georges Hospital, Hyde Park Corner, ether had been tried for dental extraction but had been abandoned due to poor results. Snow offered himself to give ether using his inhaler with such great success that he was soon appointed to give it for surgery.

James Simpson was an obstetrician in Edinburgh who took up the use of ether with enthusiasm. He gave it to women in labour to reduce their pain without making them completely insensible. He decided to investigate other volatile liquids which he tried himself at his home. He then arranged experiments with colleagues where they inhaled the liquids from glasses after dinner. Thus the anaesthetic property of chloroform was discovered and within weeks it had been used by Simpson many times.

On November 20th 1847, Simpson published 'On a New Anaesthetic Agent, More Efficient than Sulphuric Ether' in The Lancet. He said:

> *'as an inhaled anaesthetic agent, it possesses, I believe, all the advantages of sulphuric ether, without its principle disadvantages.'*

He listed these:

'a greatly less quantity of chloroform than of ether is requisite to produce the anaesthetic effect . . . its action is more rapid and complete . . . the inhalation and influence of chloroform . . . far more agreeable and pleasant than . . . ether . . . the use of chloroform will be less expensive, its perfume is not unpleasant, being required in much less quantity it is more portable . . . no special kind of inhaler or instrument is necessary.

Postscript 2. Chloroform and Ether Anaesthesia

Soon after Simpson discovered the properties of chloroform it displaced ether as the main anaesthetic agent and remained so until beyond the end of the century. Its reputation was enhanced when John Snow administered 15 minims of chloroform via a handkerchief to Queen Victoria in 1853 for the birth of Prince Leopold. This was despite his enthusiasm for ether and inhalers.

Chloroform could be administered by inhalation from a handkerchief administered from a bottle, did not require warming, was more predictable in use, did not cause nausea, induced anaesthesia more quickly, had a less unpleasant smell and could be used by the occasional operator.

Chloroform did have one disadvantage; it was soon realised that there was an increased risk of mortality with its use. It could cause ventricular fibrillation (then called cardiac syncope). At the time there was debate on whether death was due to the effects on the lungs or the heart. Attempts were made to predict which patients were at risk but it appeared to affect all ages and those in previous good health.

Writing in the British Medical Journal in October 1872 John Morgan, Surgeon and Professor from Dublin, advised that the use of ether as an alternative to chloroform should be re-examined. He said:

'ether as an agent in producing anaesthesia in surgical operations is one which may justly be considered of importance; and the claims which have been advanced on its behalf, both on the Continent and in America, where it has held its place for over a quarter of a century, deserve the attention of practitioners, who have reason, from practical experience, to distrust chloroform, or who may have found difficulty in overcoming the distrust which patients naturally feel in an agent which is accompanied by danger, and which has often instanced its subtle power by causing death under the

Facsimile of Morton's ether inhaler

Chloroform bottle for dispensing onto a handkerchief

most favourable circumstances, and where the greatest care had been taken to guard against mischance.'

Morgan quoted mortality rates from combined British and American data. He quoted a death rate of 1:23204 for ether, 1:2873 for chloroform and 1:5588 for a mixture of both.

Morgan dismissed nitrous oxide as unsuitable for anaesthesia but for the shortest procedures, and indeed it was forgotten about. However its use was revived in 1862 when Colton, whose lecture-demonstration had excited Horace Wells some 17 years earlier, again demonstrated it to dentists in Connecticut. The use of nitrous oxide was revived for dental extractions and Colton established the Colton Dental Association in New York; thousands of patients received nitrous oxide for their extractions.

In 1867 Colton visited Paris and in 1868 London where, as a result, the gas was used in the Dental Hospital; it was found to be convenient and safe for brief dental extractions. Ether was re-introduced during the 1870s and was used chiefly by those specialised in its use. However, chloroform remained popular with those who gave fewer anaesthetics because of its ease of use with an impregnated cloth or handkerchief held over the face. In 1912 the American Medical Association's Committee of Anaesthesia concluded that the use of chloroform was too risky to be justified. However it continued to be used in Britain well beyond that date.

Postscript 3. Surgery before Anaesthesia

'Cries from the Operating Theatre'

From our 21st century perspective the introduction of anaesthesia must have been a momentous advancement in the practice of surgery. However Paget did not mention anaesthesia in his memoirs or his 'Studies of Old Case Books' which he published in 1891.[1]

In his 'Case Books' he describes that many of his case records from his early surgical career were now 'useless' as 'they were made before the introduction of antiseptic surgery... improvements in nursing...ligatures and instruments, by which the risks of operation have been made comparatively trivial.' He also referred to the records of his post-mortem examinations made at St. Bartholomew's; he reported: 'these too are useless, for they were before the time of exact microscope work.' It seems odd that he should mention these advancements but make no mention of anaesthesia.

But we know it affected his wife. In 1856/57, the years anaesthesia arrived and was becoming established in England, the Pagets were living at St. Bartholomew's Hospital. Stephen Paget, their son, who published the memoirs and letters [2] wrote about his mother:-

> *'She suffered, and remembered all her life, hearing the cries from the operating-theatre a few yards off, in the years before anaesthetics-remembering him coming back and saying that she looked worse than the patient-and she always used to wonder that a day had not been set apart for national thanksgiving for the discovery of anaesthesia.'*

'Such acute pain that he fainted'

Many of the contemporary reports from surgeons cannot be relied upon to give a fully accurate picture of the horror of surgery;

perhaps they were too keen to protect the reputation of their craft and themselves; they were already under much criticism from the opponents of human dissection. They report mainly on surgical technique mentioning pain only in passing. One example was from Guy's Hospital in 1843.[3]

William Williams had a large, obviously malignant, tumour of his maxilla (upper jaw) with swelling of his face with his 'nose turned to the right', a large ulceration of the palate, with everted edges, and eye protruding from its socket. The operation was performed on 16th September 1843 and is described so:-

The patient was placed on a chair, his head supported and kept steady. An incision was made for the inner angle of the eye straight downwards towards the mouth cutting through the lip mid-way between the centre and the commissure. A second incision was made from the centre of the zygoma and carried down to the angle of the mouth, clearing the parotid duct, the integrity of which was preserved: the facial artery bled profusely and was tied. The flap was now turned upwards, and dissected away from the tumour and the maxillary bone, until the lower margin of the orbit was exposed; after dividing the infra-orbital artery and nerve which produced such acute pain, that he fainted, and the operation was suspended until he rallied. Some difficulty was experienced in clearing the edge of the orbit, and in passing the handle of the scalpel between the floor and the eye ball as the former had been raised and rendered convex from the pressure of the tumour below...

Not more than a pint of blood was lost during the operation which together with the dressing lasted nearly an hour. The parts were brought together with platinum-wire sutures; the cheek was supported internally by dossels of lint and a bread and water poultice applied externally.

He was put to bed and in about three hours vomited some blood which he had previously swallowed.

No untoward symptoms supervened to retard his recovery. The wound healed rapidly and in a few weeks he returned to his work in perfect health.

All on a conscious patient, if only it were so easy in the 21st century! Clearly we cannot rely on this as a fully accurate and unbiased case account.

'The patient is then tied up.'

Sir Charles Bell, a surgical superstar of his day, described his technique of lithotomy (removal of a stone from the bladder).[4]

> *There is a large and deep incision, low enough by the side of the anus, and the more delicate part of the operation is done, whilst yet the patient hardly feels the pain, for this piercing of the point of the scalpel is by no means so painful as the broad cut; the finer part of the operation is at once finished, whilst the patient is perfectly well.'*

The patient 'hardly feels the pain'? really! Clearly Sir Charles was the surgeon of choice if you had a bladder stone. He goes on to describe his technique for lithotripsy (breaking up the stone by crushing):-

> *'There is not the slightest difficulty in this part of the operation, neither is there pain if you do not open the instrument suddenly, and to a great extent. Take time, do it delicately and nicely, and there is neither difficulty to you nor pain to your patient.'*

Perhaps 19th century patients were clearly more stoical than today's. Here Sir Charles admits the patient feels pain but is unconcerned:-

> *'I requested the patient to tell me whether he was suffering pain. 'Oh,' he said, 'I cannot expect to get rid of a stone in the bladder without pain.' 'Nay; but tell me,' I replied, 'how do you feel-are you suffering much pain?' 'Oh, you know there must be pain.' But he never winced-never moved a muscle, never interrupted his chat,-and therefore I must presume the man was not suffering; for, if he endured pain, he must have had extraordinary fortitude neither to wince, not cry, nor even to change his voice, but readily converse with me during the operation.'*

After operating on a child, Sir Charles wrote:

> *'Nothing could be more striking, during the whole operation, than that a child so young should have so perfect a notion of the necessity of something being done for its relief, and that it should remain so submissive.'*

Possibly we are to believe that this was due to the skills of Sir Charles. However when he described the preparation for the operation of lithotomy:-

'The assistant is expected to do all the preliminary operations. See that the table be stout to bear both patient and assistant. The garters are round the wrist; the staff is introduced [from the penis through the urethra into the bladder] *the stone is felt; <u>the patient is then tied up.</u>'*

This we can believe!

Too many spectators- operation postponed until Saturday

On December 6th 1831 an operation was described in The Times not by the surgeon but by: 'A Bartholomew's Pupil'[5]

'Important Operation at St Bartholomew's Hospital'

'On Saturday last an operation was performed at St Bartholomew's Hospital, which, from its rarity, and the slight chance of success it involves, excited the greatest interest amongst the members of the faculty, and indeed, among no small number of the unprofessional public. The operation had been announced for Thursday last; but at the appointed hour a multitude so dense and intractable had assembled before the door of the operating theatre, that neither patient nor surgeon could gain admittance, and the operation was necessarily postponed to Saturday.'

Clearly surgery was a popular spectator sport. The operation was carried out by Mr Earl; it was a case of a cancer of the maxilla (upper jaw). It started with the tying of the common carotid artery in the neck in order to reduce the blood loss, cutting through the skin of the face and then removing the growth in about 4 seconds with forceps with blades shaped liked scissors. It lasted 16 minutes.

'The fortitude with which the poor creature endured the operation better deserves the name of heroism than any conduct I have ever witnessed in a patient under the knife. Indeed, she divided the admiration of the whole of the spectators with the operator, whose self-possession, serenity, and ease of manner, must be referred to that moral intrepidity which could only be acquired by long and uninterrupted devotion to acquisition of professional knowledge.

Such was the universal feeling on the subject, that when Mr Earl re-appeared in the theatre, after the patient had been removed, to make the usual comments on the case, he was received with a degree of applause that bordered on enthusiasm.'

A death on the operating table- nurses and patients on the ward shed tears

On April 11th 1831 'An eye-witness' reported in the Times[5] an 'Operation upon Hoo Loo, the Chinese, for a tumour in Guy's Hospital'. The Chinese gentleman had come to London specifically for the operation to remove his abdominal tumour. The operation was to be carried out by Mr Key. A crowd of medical, scientific and non-medical individuals had gathered to witness. 'Eye-witness' reported:-

'At 1 o'clock Sir A Cooper entered the operating theatre, and stated, that in consequence of its smallness of size, and the number who would thus be precluded witnessing the operation, it would take place in the larger anatomical theatre, whither a rush was immediately by those assembled; and although this theatre will hold nearly 1000 persons, it was crammed in every part within two minutes of the doors being opened...Hoo Loo was ushered in...with that appearance of good humour in his countenance.

The patient being laid on the table, reclining on pillows, the operation was commenced by making two elliptical incisions from the outer margin of the peduncle of the tumour to the spinous process of the pubes on each side. The knife was then carried forward along the upper part of the tumour, so as to raise a flap of integument, which was turned back. A semilunar incision was then made on each side, and two flaps of integument detached. These incisions exposed some very large subcutaneous veins, which afforded considerable haemorrhage, and were obliged to be secured by ligatures before the operation could proceed. The neck of the tumour was next detached from above, and the spermatic cords laid bare. The mass of tumour was finally dissected by a few strokes of the knife from the perineum, and the lower flaps of the integument being turned back, it became entirely detached from the patient's body.'

Alas Hoo Loo quickly died. The remainder of the account is concerned with the cause of death which was attributed to 'shock

Left: *Hoo Loo with his abdominal tumour from the Lancet 1831. <u>Above</u>: Charles Aston Key (1793-1849) Surgeon to Guy's Hospital*

to the nervous system' and the 'loss of venous blood'. It is obvious to us now that he died from surgical shock i.e. he bled to death. This did not respond to fresh air, warm applications applied to his feet and chest, fresh brandy injected into his stomach and eight ounces of blood, kindly furnished by a medical pupil, 'thrown' into the vein of the arm.

But there was no mention in the report of pain! It is clear from the report that Hoo Loo was much admired for his courage and cheerful disposition but no mention is made of the discomfort he endured. However a week later on April 19th a further report headed: 'Hoo Loo The Unfortunate Chinese' appeared in the Times. It explained:

A gentleman who was present, and understands the Chinese language, had given the following translation of the exclamations made by Hoo Loo just prior to the fainting fit before the tumour was finally removed :- 'Unleash me! Unleash me! Water! Help! Let me go.' The last articulate sounds he was heard to utter were, 'Let me be- let it remain! I can bear no more ! Unleash me!' Hoo Loo's mild and gentle manner made friends of all who knew him, and the nurses and the patients in the ward shed tears at the termination of the operation.

Desperate and ineffective methods to reduce pain

In 1795 Mr James Latta, an Edinburgh surgeon, acknowledged the pain and suffering of surgery and considered ways to alleviate this suffering to 'render the operations more tolerable than it would otherwise be'.

He said that there were two general ways in which this might be accomplished. One was by diminishing the sensibility of the patient so that he may not be capable of feeling very acute pain, the other by compressing the nerve to the part so that it cannot feel the pain so acutely. He reported:-

'Opium would perfectly answer the first intention were not its effects upon the system to be dreaded. Large doses of this medicine are very apt to bring on sickness and vomiting, which, after some operations, are much dreaded; and therefore its use is laid aside by the most judicious practitioners.'[6]

Nerve compression had been described by Mr James Moore, a London Surgeon, in 1784.[7] He lamented that 'physicians had been accused of a want of feeling for the distresses of human nature and surgeons of actual cruelty.' His clamp was designed to compress the sciatic and crural nerves so that the leg might be operated upon with decreased pain. He used it for a below knee amputation:

'at the incision of the skin he did not cry out or change a muscle of his face. At the sawing of the bones he showed marked uneasiness in his countenance but did not cry out.'

A smaller clamp was made to compress all the nerves in the axilla for operations on the arm.

Clearly attempts at reducing pain were desperate and largely ineffective. Once anaesthesia was accepted it not only reduced suffering but enabled more ambitious surgery to be carried out with increased precision and care facilitated by longer operating time.

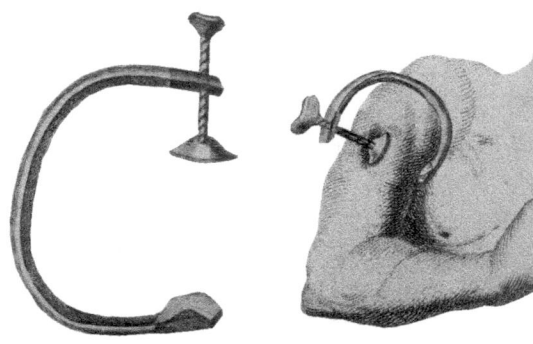
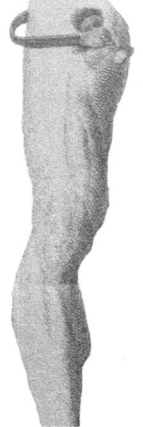

James Moore's compressor clamp and applied to axilla and leg

---

1 *Studies of Old Case Books.* Sir James Paget. 1891
2 *Memoirs and Letters of Sir James Paget.* Stephen Paget (Editor). 1908 [Note: Stephen Paget became an accomplished Surgeon himself, at the Middlesex Hospital. He was credited with describing the metastasising of malignant tumours. He wrote several books including 'Pasteur and after Pasteur' on the life of Louis Pasteur.]
3 *Guy's Hospital reports* 1843.
4 *Institutes of Surgery.* Sir Charles Bell. 1839.
5 *The Times Digital Archive*
6 *A Practical System of Surgery.* James Latta. 1795
7 *A Method of Preventing or Diminishing Pain in Several Operations of Surgery.* James Moore. 1784.

Postscript 4.   A Short Life of James Paget

Birth and Education

James Paget was born in Yarmouth on January 11th 1814. He was one of seventeen children born to his mother, nine of whom survived childhood; he had six brothers and two sisters. He was born into a prosperous household; his father was a well-respected businessman and merchant in Yarmouth. Samuel Paget was a brewery and a ship owner but he had built most of his wealth by supplying provisions to the North Sea fleet in Yarmouth by contract with the Admiralty. He was to lose his fortune later in life and embrace poverty.

The young James had a modest education at one of the two boys' schools in Yarmouth. It was kept by a Mr. Bowles, a minister at the chapel and a former actor. At the age of thirteen James might have been expecting to progress his education at the public school Charterhouse, as his three older brothers had, but by this time his father had fallen on hard times in business and could not afford the expense.

James continued at Mr. Bowles's school until he was sixteen and was unable to obtain a better education. He described that at Mr Bowles's he acquired such knowledge as he could but not what he called 'classic knowledge' so that he was unable to take part in learned table talk except as a listener. He thus was unable to acquire the fitness or facility for social life which was attributed to those who had enjoyed the advantage of public schools and universities.

The young James flirted with the idea of a career in the Navy, having seen the smartly dressed young officers who came ashore at Yarmouth and become envious of the attention paid to them, particularly by ladies. An advantage of this career was that it was a profession for gentlemen that could be entered without great expense. However James's mother did not wish him to leave home for officers' training so he stayed in Yarmouth for a career in the medical profession.

Apprentice to a surgeon in Yarmouth 1830-1834

In 1830, at the age of sixteen, James became an apprentice to Mr. Charles Costerton in Yarmouth, a 'surgeon'- meaning a general practitioner. The apprenticeship 'to learn the art and mystery of a Surgeon and Apothecary' was to last for five years as required by the Society of Apothecaries but after four and a half he would be allowed to go to London for hospital based study. Somehow the premium of one hundred guineas was found for him to enter the apprenticeship.

James considered that five years' apprenticeship was too long a training for the job, but he considered he received 'a good grounding in dispensing, a practical knowledge of medicines and of making them, account keeping, business habits for practice and care, neatness and cleanliness in minor surgery.'

Much of his work as an apprentice was dull and tedious and considered at times useless by him. Each day at the surgery he would dispense, see out-patients of the poorer classes, receive messages, make appointments, keep accounts, make out bills and receive payments. He would write notes for his master from his dictation, as well as prescriptions, which he made up, and sent in boxes as well as leeches. Initially his teacher was one of the more senior apprentices.

The out-patient work that he described consisted of patients with ulcerated legs (which had to be bandaged), coughs, colds, occasionally slight injuries and chiefly women who came to be bled once or twice a year particularly when they came to Yarmouth for market days. It was felt at the time that occasional bleeding was a help to health. James concluded that he considered that neither good nor harm ever came from this practice.

James was taught anatomy by Mr. Costerton who had been a pupil at St. Bartholomew's Hospital. He was able to dissect some internal organs and amputated limbs and attended lectures on bones given in The Angel Inn at Acle by Mr. Randall, a young surgeon. James was able to observe operations carried out by his master and other surgeons of the town. His first case notes were written about a young boatman who was injured on the 17th February 1830 (before his apprenticeship) by an accidental gunshot injury which required the amputation of an arm and a leg. The operation was carried out without anaesthetic. He read avidly books on medicines and lectures published in the Lancet by Abernathy, Astley Cooper and Lawrence.

During his time as an apprentice he took an avid interest in the study of botany which 'introduced him into the society of studious and observant men' and encouraged the habit of observing and helped him 'learn the habits of orderly arrangement'. He described the knowledge he had gained from his study of botany as useless, but the discipline he gained from acquiring it he felt was 'beyond all price'.

Pupillage at St. Bartholomew's Hospital 1834-1835

James finished his apprenticeship early and in October 1834 transferred to St. Bartholomew's Hospital to finish his training. He had never previously been more than twenty miles from his home. He was able to afford the one hundred guineas fees by borrowing it from his brother as his father was no longer in a position to be of any financial help to him.

He found the hospital to be disorganised and many of those who were studying there were idle. Most had been to university and James observed that they felt that their university education could give them 'sufficient claim to as much success as they cared for'. Work had to be self-guided as there was little help for the students from their teachers. The school was declining in numbers and fitness for teaching although it offered opportunities enough for those who could teach themselves with little or no personal guidance.

There was a small library from which books were given out, but there was no reading room. Several of the pupils formed a club from a small room over a baker's shop nearby where they would sit and read or play cards between work. The 'dead house' was a shed which was damp and dirty. Post-mortem examinations were made on a table surrounded by pupils; examinations were carried out in a rough manner with little instruction and nothing was really carefully examined or taught and pathology hardly considered. James describes the museum as being in good order and later he became a curator, using his skills in studying organisation that he had gained from his study of botany.

There were lectures at the hospital for the pupils; he described some as good and others not. However, he was not a regular attender. He found that he was able to learn more from reading, dissecting and working in the dead house and the out-patients' room. It is clear that James had a sharp intellect and his reading rewarded him with being the best anatomist in his year. In this he was advantaged as he had studied French and German. Germany at the time was at the forefront of medical knowledge and

Bart's Hospital early 19th century

innovation and therefore James was able to read German medical and anatomy books fluently.

In his first year at Bart's James discovered the nematode, trichinella spiralis, which causes the disease Trichinosis. He claimed that everyone in the dissecting room saw little specks in the muscles of the cadavers but James was curious about them. He felt that his power of observation had been conditioned by his botany studies and he wished to study the parasite with the aid of a microscope. St Bartholomew's Hospital at the time had no microscope so he approached the British Museum knowing there was a microscope there in the Botany department. He made drawings of the parasite and presented his findings to the Abernethy Society. His specimens were taken to the biologist Sir Richard Owen who named the parasite and presented a paper to the Zoological Society and was thus widely accredited with James's discovery (showing that academics seeking advancement from the work of others were ever present). However Paget was pleased that his work enhanced his position within the hospital and made him contacts to be useful in future studies.

Later he attended the wards and the out-patient room. The opportunity for practical study was very little. A dressership cost ten guineas a year, which he could not afford, and he believed offered less experience than he had previously achieved in Yarmouth. There was no active teaching on the wards other than by looking on and by occasional casual talking about cases.

In his second year Paget passed his examination for the membership of the College of Surgeons (the Fellowship of the College of Surgeons had not yet been instituted.) In order to take the examination he had to complete eighteen months of study at the hospital and two and a half years of study elsewhere. He found the examination to be very simple; he was now legally qualified to practise.

Waiting for suitable employment 1836-43

Following the year of his qualification James spent seven years in London awaiting a suitable hospital appointment. He did not wish to become engaged in rural practice.

In order to make ends meet he took pupils, the first of whom lodged with him. He spent three months studying in Paris. He preferred writing to teaching and became a sub-editor of the Medical Gazette, often writing translations from French and German papers, which earned him between £50 to £70 per year.

It was during these years that he became curator of the museum at Bart's. This work was paid but required attendance for long hours for menial work but it led to better things. Eventually it led to a Demonstratorship in morbid anatomy at Bart's and to employment to re-write the Pathological Catalogue of the College of Surgeons Museum which led to the College Professorship. During these years he became engaged to be married to a Miss North.

Teaching appointments at Bart's Hospital 1843-1851

In 1843 Paget was appointed to a lectureship in physiology and to a separate appointment, a lectureship in anatomy. Bart's school had fallen to its lowest level for years with only thirty six students having entered for the lectures in anatomy. It was decided at this time that the school should become residential; previously the students had lived outside the hospital, and James Paget became the first Warden to the pupils who were to be the first residents.

As Warden he was responsible for the general supervision and guidance of the students as well as 'discipline in respect of hours of coming in at night, behaviour in hall, the control and supply of wine, the breaking up of noisy parties and occasional expulsion for gross misconduct'.

He lived in the hospital for eight years between 1843 and 1851. In 1847 to his duties at Bart's were added the Professorship of Anatomy and Surgery at the College of Surgeons which he held for

Paget lecturing at St Bart's on the problems caused by unsuitable footwear. From 'Pictorial World' July 25th 1874

six years and he became the Assistant Surgeon at the hospital in 1846. He finished the catalogue of the hospital museum specimens and in 1849 he finished the pathology catalogue at the College of Surgeons. Once appointed to the hospital and taking up the position of Warden he now enjoyed life after seven years of loneliness. He lived in and dined in the hospital each day with senior students with excellent and varied food and conversation about hospital affairs, science and news. He married Miss North in May 1844.

In 1843 the Royal College of Surgeons instituted its fellowship. James Paget became one of the three hundred first fellows of the college which were appointed at that stage by election and in 1851 he was elected as a Fellow of the Royal Society.

While he was lecturing in physiology at Bart's one of his pupils was Kirkes. Kirkes subsequently wrote a book on physiology which became the standard textbook on the subject for many decades and went to many editions. Sir Michael Foster asserts that Kirkes's Handbook of Physiology first edition was wholly based on Paget's lectures and the book followed those lectures in all of its earlier editions.

Private Practice 1851 onwards

After seven years as Warden at Bart's he described that he was getting older, whereas the pupils in succession were not and the maintenance of the rules was becoming tedious and the noisy parties with singing late at night and the 'chaffing' of people in the street became 'intolerable'. He also found that his income of five to six hundred pounds per year was becoming insufficient for the comfort and the training of his children who were increasing in number.

In 1851 he gave up the wardenship of the hospital and moved into 24 Henrietta Street, to set up his own home with his wife and children and to start his own surgical practice.

Once he started in private practice James Paget describes his experience up to that time in practical surgery as limited but not deficient. He had been Assistant Surgeon at St. Bartholomew's for four years, mainly doing the duties of others in the out-patients room and wards. He considered himself more fit for practice than some others. He felt that he was not disqualified from success by the hindrances which had spoilt the careers of many others who were 'idle, time wasting, unbusinesslike, unpunctual or unable or unwilling to work with others'.

His practice gradually increased to become the most lucrative in London, earning him over ten thousand pounds per year. As well as wealth he accumulated the largest number of interesting cases and benefited from positive relationships with others in the profession with whom he cooperated on cases. He was frequently called upon by other practitioners to give his opinion or accept difficult cases from them. Other medical men would attend his house with their patients.

He observed that his practice increased when he attended cases that were much talked about. When he attended the Princess of Wales the increase in his practice the following year was twice as much as previously. In 1858 he moved from Henrietta Street, the lease having expired, to Harewood Place, and this coincided with a significant increase in his income.

He now remained the Surgeon to the hospital and a joint lecturer in surgery. He attended the hospital on at least six days a week as well as for urgent cases at any time of the day or night. In the mornings he would operate on private patients and visit between eight and ten important cases. He would then see consultations at his home, maybe fifteen to twenty cases in three hours, and then visit private patients in distant parts of the town. After dinner would come letter writing, book-keeping and reading or other work going into the night. Sometimes his routine work would be interrupted by having to visit patients out of town by train. He estimated that he usually worked about sixteen hours per day.

In 1871 he was ill for three months and resigned as Surgeon to the hospital. Soon after he became Sir James Paget, having been conferred a baronetcy (a hereditary title of higher rank than Knight). At that stage he gave up operating but continued with other work; his income decreased but he could afford it.

Medical Politics

It would not be possible for someone of Paget's ability and influence to avoid Medical Politics. He was elected to the council of the Royal College of Surgeons in 1865, the senate of the University of London in 1860 and a member of the Medical Council in 1876. He became Vice Chancellor of the university in 1883.

He felt that medical political work produced 'less good done, in proportion to the time spent, than in anything in which I have been engaged'.

Controversy

In 1854 Paget caused controversy when he became an examiner for the entry of surgeons into the employment of the East India Company. Paget examined surgery; there were also examiners in anatomy, medicine and elementary sciences. In order to be eligible for the examination a surgeon had to be a member of one of the three colleges of surgeons.

Sir Joseph Hooker wrote an account of these examinations, which took place two weeks annually. He described that the examiners all met at Paget's house for discussing the papers that were set and for the final examination of the answers. Paget required every candidate to carry out a major and a minor operation on a dead subject. Hooker described that 'the ignorance and incapacity of a very large proportion of the few first batches of candidates was astounding; very many could not tell the freezing and boiling points of water'.

Paget failed so many of the candidates that the positions with the East India Company were not filled. The colleges of surgeons were furious as by failing the candidates Paget had pronounced unsuccessful candidates to be incompetent even though they had passed the examinations to become members of the surgical colleges.

Old Age

In September 1893 he moved to 5 Park Square West, Regents Park. He continued to see a few patients who came to him. In his old age he enjoyed novel reading in both French and English. He particularly liked the works of George Eliot, whom he befriended, and who sent him her books as well as Tennyson. He died on the night of December 30th 1899.

Appendix: List of chief writings of James Paget

1834 *Sketch of the Natural History of Yarmouth and its Neighbourhood.* By C. J. and James Paget. F. Skill: Yarmouth. 8vo. Pp. 88.

1835 *Account of the Trichina Spiralis.* Trans. Abernethian Society.

1840 *On White Spots on the surface of the Heart, and on the frequency of Pericarditis.* Trans. Med. Chir. Soc., xxiii 29.

1842 *On the chief results obtained by the use of the Microscope in the study of Human Anatomy and Physiology.* British and Foreign Medical Review.

On the relation between the Symmetry and the Diseases of the Body. Trans. Med. Chir. Soc., xxv. 30.

1844 *Report on the Progress of Human Anatomy and Physiology in the year 1842-3.* British and Foreign Medical Review.

On Obstructions of the Branches of the Pulmonary Artery. Trans.Med. Chir. Soc., xxvii. 162.

Examination of a Cyst containing seminal fluid. Ibid. 398.

1845 *Report on the Progress of Human Anatomy and Physiology in the year 1843-4.* British and Foreign Medical Review.

Address at the Abernethian Society (Fiftieth Session).

Additional Observations on Obstructions of the Pulmonary Arteries. Trans. Med. Chir. Soc, xxv111. 353.

1846 *Catalogue of the Pathological Specimens in the Museum of St.Bartholomew's Hospital.* Pp. 487.

First Volume of the Pathological Catalogue of the College of Surgeons Museum Pp. 144

Report on the Progress of Human Anatomy and Physiology during the year 1844-5. British and Foreign Medical Review.

Records of Harvey. in extracts from the Journals of the Royal Hospital of St. Bartholomew. London : John Churchill. Pp. 37.

(Reprinted in the St. Bartholomew's Hospital Reports, xxii. 1886)

On the Motives to Industry in the Study of Medicine : an address at the opening of the Hospital-session, October, 1846.

Account of a case in which the corpus callosum, fornix, and septum lucidum were imperfectly formed. Trans. Med. Chir. Soc, xxix. 55.

1847 *Second Volume of the College Catalogue.* Pp. 255.

Lectures on Nutrition (Arris and Gale Lectures). Med. Times.

1848 *Third Volume of the College Catalogue.* Pp. 287.

Handbook of Physiology. By William Senhouse Kirkes and James Paget.

1848 *Lectures on the Life of the Blood* (Arris and Gale Lectures). Med.Times.

Account of a dislocation in consequence of disease of the first and second cervical vertebrae. Trans. Med. Chir. Soc., xxxi. 285.

1849 *Fourth and Fifth Volumes of the College Catalogue.* Pp. 350 + 182.

Lectures on the Processes of Repair and Reproduction after Injury (Arris and Gale Lectures). Med. Times.

1850 *Lectures on Inflammation* (Arris and Gale Lectures). Med. Times.

On the Freezing of the Albumen of Eggs. Phil. Trans. Roy. Soc.

On Fatty Degeneration of the Small Blood-vessels of the Brain, and its relation to Apoplexy. Trans. Abern. Soc.

A case of Aneurismal Dilatation of the Popliteal Artery, treated with Pressure. Ibid.

1851 *Lectures on Tumours* (Arris and Gale Lectures). Med. Times and Gazette.

On the Recent Progress of Anatomy, and its influence on Surgery. A lecture at the College of Surgeons, July 2. Med. Times and Gazette.

1852 *Lectures on Malignant Tumours* (Arris and Gale Lectures). Med, Times and Gazette.

1853 *First Edition of Lectures on Surgical Pathology*: being the Arris and Gale Lectures, with additions. Two volumes. Pp. 499 + 637. Longmans.

Two cases of inguinal hernia, in which the sac was pushed back with the intestines. Med. Times and Gazette.

1854 *On the importance of the study of Physiology, as a branch of education. for all classes.* Lecture at the Royal Institution, June.

Account of a growth of cartilage in a testicle and its lymphatics, and in other parts. Trans. Med. Chir. Soc., xxxviii. 247.

1856 *The Physiognomy of the Human Mind.* Quarterly Review, September.

1857 *On the Cause of the Rhythmic Motion of the Heart : the Groonian Lecture.* Proc. Roy. Soc.

On the hereditary transmission of tendencies to cancerous and other tumours. Med. Times and Gazette.

Account of a case in which the administration of chloroform was fatal. Ibid.

1858 *Notes of practice among the Out-patients of St. Bartholomew's Hospital.* Med. Times and Gazette.

A case of aneurism of the external iliac and femoral arteries. Ibid.

1859 *The Chronometry of Life : a Lecture at the Royal Institution.* Med. Times and Gazette.

1860 *Articles in Holmes's System of Surgery.*

1862 *On the treatment of patients after surgical operations : the Address in Surgery at the annual meeting of the British Medical Association.* Med. Times and Gazette.

1863 *Second Edition* (with Sir W. Turner) *of the Lectures on Surgical Pathology.*

Introductory Address at the opening of the Hospital session. Lancet, ii. 1863.

1864 *Scarlet fever after operations.* British Medical Journal, ii. 1864.

1865 *Inaugural Address at the opening of the new buildings of the Leeds School of Medicine.*

Cases of Chronic Pyaemia. St. Bartholomew's Hospital Reports, i. 1.

1866 *Gouty and some other forms of Phlebitis.* St. B. H. Reports, ii 82.

1867 *Senile Scrofula.* St. B. H. Reports, iii. 412.

The Various Risks of Operations. Lancet, ii. 1867.

Cases that Bone-setters cure. B.M.J., ii. 1867.

1868 *The Calamities of Surgery: a clinical lecture,* 1868.

Stammering with other organs than those of speech. B.M.J., ii. 1868.

1869 *Presidential Address to the Clinical Society of London.*

What becomes of Medical Students. St. B. H. Reports, v. 238.

Residual Abscesses. Ibid., v. 73.

Treatment of Carbuncle. Lancet, i. 1869.

A case of suppression of urine very slowly fatal. Trans. Clin. Soc., ii.

1870 *Third Edition of the Lectures on Surgical Pathology.*

Sexual Hypochondriasis : a clinical lecture.

The production of some of the loose bodies in joints. St. B. Reports, vi. 1.

Cancer following ichthyosis of the tongue. Clin. Soc., iii.

Necrosis of the femur without external inflammation. Ibid.

Wasting of part of the tongue in connection with disease of the occipital bone. Ibid.

1871 *On dissection-wounds.* Lancet, i. 1871.

Cancerous tumours of bone. Trans. Med. Chir. Soc., liv.

A case illustrating certain nervous disorders. St. B. H. Reports, vii. 67.

1872 *Lectures on Strangulated Hernia.* B.M.J., i. and ii., 1872.

1873 *Memoir of William and Edward Ormerod.* St. B. H. Reports, ix.

Lectures on Nervous Mimicry. Lancet, ii. 1873.

1874 *Disease of the Mammary Areola preceding Cancer of the Gland.* St. B. H. Reports, x. 87.

Presidential Address in the Section of Surgery, annual meeting of the British Medical Association. B.M.J., ii. 1874.

Remarks on Pyaemia. Trans. Clin. Soc., vii.

Remarks on Cancer. Trans. Path. Soc., xxv.

1875 *First Edition of Clinical Lectures and Essays.* Edited with notes by Mr. Howard Marsh. Longmans.

1876 *Fourth Edition of the Lectures on Surgical Pathology.*

On a form of chronic inflammation of the bones (osteitis deformans). Trans. Med. Chir. Soc., lx

On some of the sequels of Typhoid Fever. St. B. H. Reports, xii. 1.

On certain points in the pathology of Syphilis. B.M.J., i. 1876.

Presidential Address to the Royal Medical and Chirurgical Society of London. Ibid.

1877 *Hunterian Oration,* B.M.J., i. 1877.

Presidential Address, Roy. Med. Chir. Society.

Cases of Branchial Fistulae in the External Ears. Trans. Med. Chir. Soc., lxi. '

1878 *On Indurations of the Breast becoming cancerous.* St.B. H. Reports,xiv. 65.

The Contrast of Temperance with Abstinence. Contemporary Review, November.

1879 *Second Edition of the Clinical Lectures and Essays; including four lectures on Gout in some of its surgical relations.*

Memoir of George William Callender. St. B. H. Reports, xv.

Anaesthetics : the History of a Discovery. Nineteenth Century.

Case of polypi of the antrum. Trans. Clin. Soc., xii.

1880 *Elemental Pathology: the Presidential Address in the Section of Pathology, at the annual meeting of the British Medical Association.* B.M.J., ii. 1880.

Suggestions for the making of Pathological Catalogues. Ibid.

Theology and Science : an address at the Leeds Clergy School, December, 1880.

1881 *Presidential Address at the International Medical Congress in London.*

The Vivisection Question. Nineteenth Century.

First Volume of the new edition of the College Catalogue (with Dr. Goodhart and Mr. Alban Doran).

1882 *On some Rare and New Diseases: the Bradshawe Lecture at theCollege of Surgeons.* B.M.J., 1882.

Notes on seven additional cases of Osteitis Deformans. Trans. Med. Chir. Soc., lxv.

1883 *Second Volume of the new edition of the College Catalogue.*

Address on the Collective Investigation of Disease. B.M.J., ii. 1882.

1884 *Third Volume of the new edition of the College Catalogue.*

On the National Value of Public Health: an address in connection with the International Health Exhibition. B.M.J., i. 1884.

1885 *Fourth Volume of the new edition of the College Catalogue.*

Remarks on Charcot's Disease. Trans. Clin. Soc., xviii.

An Address at Netley Hospital, at the presentation of prizes.

St. Bartholornew's Hospital and School, fifty years ago : an address to the Abernethian Society. B.M.J., i. 1885.

1886 *An address at Oxford, on the unveiling of John Hunter's statue in the University Museum.* B.M.J., i. 1886.

1887 *On Cancer and Cancerous Diseases : the Morton Lecture at the College of Surgeons.* B.M.J., ii. 1887.

On the Future of Pathology: Presidential Address to the Pathological Society of London. Trans. Path. Soc., xxxviii.

Report on Charcot's Disease (with other members of the Clinical Society's Committee). Trans. Olin. Soc., xx.

Report of the Pasteur Commission (with Mr. Victor Horsley).

On the utility of scientific work in practice : an introductory address at Owens College. B.M.J., ii. 1887.

Memoir of Sir George Barrows. St. B. H. Reports, xxiii.

1888 *Address to London University Extension Students.*

1889 *Address at the Mansion House meeting in recognition of the Pasteur treatment against rabies.*

1890 *Address at University College, Liverpool.*

1891 *Studies of Old Case-books : seventeen essays on subjects in surgical pathology and practice.* Longmans, 1891. Pp. 168.

A short account of M. Pasteur's work. Nature.

A short paper in the Virchow Festschrift.

1894 *An address to the Abernethian Society, at the beginning of its hundredth session.* B.M.J., ii. 1894.

Further Reading

Richard Gordon. *The Sleep of Life - a Novel.*

Novelist and former anaesthetist Gordon Ostlere, writing as Richard Gordon, became most famous for his 'Doctor' books which were widely read and filmed in the 1960s and 70s. Although these may seem dated now this well researched and skilfully written account of the introduction of anaesthesia into London, in novel form, is as fresh now as when it was written in 1975. The story of William Morton's attempt to disguise ether and patent his invention 'Letheon' to prosper financially entertains as much as it informs us of the history and context of the time. Although out of print there are numerous copies to be bought on-line.

Stephanie J Snow. *Blessed Days of Anaesthesia - How Anaesthetics Changed the World.*

This book relates the history of anaesthesia and its social history and attitudes to pain from the mid eighteenth century and the discovery of anaesthesia into the twentieth. The book is extensively referenced with a large suggested reading list.

Peter Stanley. *For Fear of Pain. British Surgery, 1790-1850.*

This is a well written an extensively referenced history of surgery in Britain up to the introduction and acceptance of anaesthesia. This book is eminently readable and suitable for both the interested social reader and the historian looking for an overview of the subject.

Acknowledgements

'Escape from Pain' was initially published in the monthly periodical 'Nineteenth Century'.

The Chapter on the life of Paget and the list of his writings was principally sourced from 'Memories and Letters of Sir James Paget' edited by his son Simon Paget, published in 1903.

I acknowledge the Libraries of the Royal Society of Medicine, the Royal College of Surgeons as well as the British Library as the sources of much of my background reading. The Times digital archive was accessed online via Lincolnshire Libraries.

The images were sourced as follows: Cartoon from Punch: John Rylands Library, Manchester. St Bartholomew's Hospital and Paget lecturing : St. Bartholomew's Hospital Archive. Horace Wells statue: self. William Morton administering ether and Charles Key: US National Library of Medicine. Hoo Loo: Lancet 1871. James Moore's compressor clamp: James Latta's Practical system if Surgery 1795. Other images: Wellcome Library, London.

Andrew Sadler December 2014

www.ingramcontent.com/pod-product-compliance
Lightning Source LLC
Chambersburg PA
CBHW072246170526
45158CB00003BA/1016